JP's Classic Work on the
Institutionalization of
The Retail Pharmacist

JP's 20 Simple Rules For the Successful and Satisfying Practice of Pharmacy

By:

Jim Plagakis, R.Ph.

ISBN: 978-1-257-01683-9

Jim Plagakis, R.Ph.

For more than twenty years Jim Plagakis has been the author of the popular Drug Topics magazine column "JP at Large". He has been a prescient observer of the drug store industry and his predictions have been consistently accurate. Jim loves pharmacy and he knows that it is the job that can present problems. The profession is just fine.

"My method is to take the utmost trouble to find the right thing to say, and then to say it with the utmost levity"
George Bernard Shaw

This is the third edition of my classic examination of the institutionalization of the modern retail pharmacist. It was written in 2006 and was a desk top publication that had modest sales. Lulu.com is a better publication choice. The presentation is superior to desktop and top quality. I have removed all of the superfluous commentary that I added to the second edition. I added a few sentences. What you are getting here is the virgin text of the original, untouched and, in many spots, a bit over the top. Rules Number One and Number Twenty are the most important things a pharmacist can do for themselves.

Author's Note

The file for the first edition was lost when my entire Documents File was deleted in 2008 during a perfect storm of inattentiveness. I will never be able to print additional copies. Hold on to your copy. You may want me to write something in it someday.

The second edition is not a virgin. It is all gussied up with catchy commentary and ready to take her love to town. It is not for sale on my website. If you would like an autographed copy, contact me at

jpgakis@hotmail.com

The Cardinal Rule
Number One
Always Put Yourself First

This is an example of the old <u>you can drag a pharmacist out of the drug store, but you can't make her relax.</u> The last thing on your mind is putting yourself first. That would be selfish. The morning technician is working overtime. Your partner cannot miss another dental appointment. You have doctors to call and another order to complete.

Mrs. McKeever is standing at the counter with her hands on her hips. She is taking it out on you and you didn't do anything.

"Mrs. McKeever. Your insurance company denied the Enbrel. I'm sorry." You mean it. You do feel sorry. Your stomach is burning again.

"You promised." Mrs. McKeever said angrily.

"I said that I'd try," you tell her. "I did try and they said No." She has pain, but there is nothing you can do.

The store manager wants to talk about your vacation request. You had asked for two weeks in July. You have promised the kids a vacation out west. A week in Yosemite and a week on a California beach. You have worked for this company for ten years and have never had a vacation in the

This is an example of the old <u>you can drag a pharmacist out of the drug store, but you can't make her relax.</u> The last thing on your mind is putting yourself first. That would be selfish. The morning technician is working overtime. Your partner cannot miss another dental appointment. You have doctors to call and another order to complete.

Mrs. McKeever is standing at the counter with her hands on her hips. She is taking it out on you and you didn't do anything.

"Mrs. McKeever. Your insurance company denied the Enbrel. I'm sorry." You mean it. You do feel sorry. Your stomach is burning again.

"You promised." Mrs. McKeever said angrily.

"I said that I'd try," you tell her. "I did try and they said No." She has pain, but there is nothing you can do.

The store manager wants to talk about your vacation request. You had asked for two weeks in July. You have promised the kids a vacation out west. A week in Yosemite and a week on a California beach. You have worked for this company for ten years and have never had a vacation in the summer.

The store manager is about to deny your request. You can smell it coming and that makes your stomach burn even more. He's going to tell you to reschedule for next October. He'll tell you to take the kids out of school. Again. You husband has had the two weeks in July marked in his calendar since February.

This job is making you sick. You take omeprazole 20 mg morning and night and space a 300 mg ranitidine in between and you still need chugs of Maalox on days like this. You deserve the vacation in July. You earned it.

How can you put yourself first? This job, by its very nature, is not designed to let any pharmacist thrive and that is our own fault. Pharmacists have not been the drivers of the job of working in a pharmacy for at least three decades. We have quietly swallowed indignities and insults and have barely made a whimper. You must take care of yourself.

If you let it, your *job* as a pharmacist will grind you down and spit you out, but I really do not have to tell you that, do I. I am just a reminder. A mnemonic device, if you will. They play the national anthem before a baseball game to remind you of patriotism. Put this book right beside the mirror you use in the morning to remind you that you are the reason for every season. Pharmacy still belongs to you.

Your job can affect your blood pressure, your heart rhythm and your regularity. You may have fitful nights with bizarre dreams. Cold sweats drench you. You spend the entire night dealing with problems with a prescription, over and over again and never getting it right.

Your sex life can be affected. You need a release and your wife wants nothing to do with the mean man who got home ready to rumble. You get home at nine thirty at night and demand of your spouse, "Where's my dinner? I work my ass off for you and the kids and you can't even have a decent meal ready for me."

Your partner just walks away. She turns her back so you can't see her stricken face. She wipes her eyes. You say you are sorry, but she brushes your hand from her arm.

She can no longer put up with this behavior. Last night, you were not hungry. You took a beer in to watch the game. Then a second beer, and a third. You fell asleep, fully dressed, in your chair. This is hunkering down to survive. Not even close to putting yourself first.

Notice that I am talking about the *job*. Not the *profession*. You have made the mistake of thinking that they are one and the same. Pharmacy is an impressive profession. It is the *job* that can ruin you.

Don't even think about your family. Your real life is not even in play here. Your job is what you do to make a living to support and enhance your real life. Your real life can be wrecked by your inability to take care of yourself.

I was institutionalized. Dignity was not in play for me. I thought about it. I talked about it. I wrote about it. I could get dignity by putting myself first. I did not know how to do it.

You are an intelligent, well-trained professional. You have the knowledge and experience that earns you a six figure income. You can have it your way. You just don't know it.

You have to learn how to say NO. Unfortunately, you are on your own. One good rule is to consider the situation and determine if they are about to take advantage of you… one more time. You will learn to spot them coming. They will no longer be able to sneak up on you.

Talk about it with someone who cares about you. Take baby steps. You'll stumble, but you won't go back. Learn to **put yourself first.** If not now, when? Your life depends on it.

Rule Number Two
Never Dip Your Pen In Company Ink

Let your imagination run. The young woman in the illustration below is a pharmacy manager for a Big Box store. The guy is a new pharmacy technician. They work together a lot and .....

© www.danheller.com

Rule Number Two
Never Dip Your Pen in Company ink

This is an easy rule to understand, but it may be the hardest to follow. The hunger of the organs for each other. You know exactly what I am talking about.

Doctor & Nurse. Lawyer & Secretary. Pharmacist & Technician. What could be more natural? You are close together 8 hours a day. 5 days a week. Hormones are flowing. You are too busy and exhausted at the end of your day. Date? Who has the time and energy? The pharmacist's single life with money does not match the light beer commercials.

Put the two of you together. Add a little loneliness, a few *innocent* touches during the day. Secretive glances that end in a shared meal at that little candlelit Italian place around the corner. It was a hard day, you tell yourself. You earned this and Steve is a good man.

This can start out very innocently. You both know instinctively that there is danger here and that makes it even better. You ask the waiter for a bottle of Harazthy Vineyards Zinfandel. You smile at Steve and touch his hand, "Nothing can happen, Steve. It could make things complicated at work."

You know each other very well and you like each other a lot. You can be adults and share a meal after work. You avoid shop talk. The meal is very good. You order a second bottle of wine. You dance a little. Then comes the slow, deep-tone music. You look at each other. Why not?

What's wrong with a little hug in the parking lot? Why can't you share a little kiss, just one?"

Next thing you know *you are dipping your pen in company ink.* That is an inelegant metaphor from the days when women pharmacists were uncommon.

Steve spends the night at your place. You are both worn out from lovemaking and tenderness. Enervated from the excitement. You sleep well in each other's arms. You look at his peaceful face in sleep. This is good, you think,. but it is best that it not happen again.

Oh gawd, I fall down and hurt myself laughing. *We are never going to let that happen again.* That's when you both say: *Let's be adults about this.* You both nod looking real serious, but you know that you want it again. There is something mouth-watering delicious about forbidden love.

The inelegant *company ink* metaphor is outdated. It was passed on to me by a wise old pharmacist who had found out the hard way. What did I do? I found out the hard way and I did it twice. The first technician I slept with became my second wife and my second failed marriage. The second one damn near did allegorical acrobatics in her attempt to become my third wife.

Neither worked out. The second one did more than innocent touching as we worked on prescriptions. This was in the 1970s in lusty, swinging San Diego. In the 21st Century, it could be called sexual harassment. She was the harasser and I was putty in her hands. She had a thing she did with her knee from behind me as I worked.

Eventually, she started to expect favored treatment. Off every Saturday and Sunday. She wanted to work only the early shift. The other tech knew who was doing what to whom. She had a justifiable fit one day when my tech/mistress and I came out of the bathroom together. one afternoon. What was I thinking? This was a pharmacy, not a strip club. I tried to make it better, but tech number two demanded a transfer over the hill to La Mesa. I did not send her written request to Los Angeles.

That was a good thing because my mistress had a surprise for me one afternoon. We were in bed. After. She had her legs tangled up with mine. I remember the smells, the sweat. I was smoking a cigarette with my eyes closed and then she dropped it on me. She demanded a raise. My only thought was: *What the hell have I gotten myself into.*

Her demands got so over-the-top that I had to fire her. She said that she would ruin me and did her best in multiple letters to the Pharmacy District Manager. He came with the letters and a *oh-you-sly-fox* grin on his face. "What the hell were you thinking?" Then, "I should be so lucky." This was 1977 in San Diego. A place of beautiful hard bodies and soft morals.

You girls do not escape. There is big danger in a boy toy.

What if you are married? You have the same hormones, the same closeness with the tech. You share stresses that your spouse can't understand. There is a lot to lose.

Some of you could tell me your racy stories. This is not easy territory to navigate.

Rule Number Three
They are Impaired, Stupid
Don't Piss Them Off

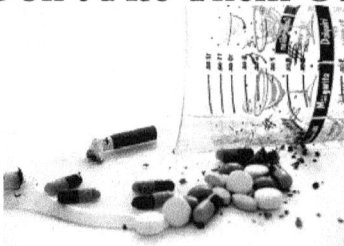

My cousin's wife is a pharmacist in Ohio. Thirty years ago, she was near the end of a long work day. She was tired and impatient. She took it out on another of a line of drug seekers that evening. She had run out of her day's supply of tact, friendliness and common courtesy. She treated this large, desperate man poorly and she paid a price.

Jane did her job properly and her reward was having to have the police meet her at closing time so she could walk to her car safely. All she did was to refuse to refill a prescription for a controlled substance because the Rx was for a 30 days supply and it had only been 5 days.

The police? Why did she need the police? She needed protection because she really pissed this guy off. She broke all of the rules. She showed no consideration to the drug seekers story even if it was a lie. She blew him off. She turned her back on him and walked away while he was arguing. She turned and mouthed off at him. Something like, "Yeah, yeah, yeah. I hear your pathetic story. I'm still not filling the prescription."

What did Jane do wrong? Just about everything. She had a perfect storm of inattention because she was worn out from a long and stressful day. This patient was a demanding and abusive drug seeker, but he was still a human being in need. She should have stopped for a few seconds, taken a deep breath and look at him as a human being who needed some help. She was rude and dismissive. She was just happy to get rid of him.

Then she really blew it. She made it worse when another pharmacy called for a transfer for the prescription in question and she made sure that the other pharmacy would refuse him too.

The patient became dangerous. He was waiting for her that night. She had to run for her car, threatening him with her canister of Mace. She made arrangements for the police to meet her at the door every day she worked the late shift. Eventually, the police quit coming. Then she quit a very good job.

I will never suggest that you pander to drug seekers. I just want you to be aware of the dangers and to watch your own back. You can refuse and still be pleasant. You can show some concern. In the end, you can refuse because *Your hands are tied. It is the law.* Drug seekers can be dangerous.

I am sure that you have seen plenty of patients with that droopy eye stare across the counter. "I want my Vicodins, my

Somas and my Xanaxs right now." They are very good at this. "My aunt died. I'm leaving today for the funeral."

Well now, that's quite a story. I ask quick questions. *Where is the funeral? Driving or flying? What airline? Your Mom's sister? What's your aunt's name?* If there is any hesitation, I suggest that she have a drug store in Albuquerque call for a transfer when her supply gets lower.

Most of all, I try to be courteous. I am not the arbiter of how these people live their lives. I call them *Maam and Sir.* I use first names only if I know I have their permission. I have had plenty of talks with benzodiazepine patients. I give them all of the withdrawal information I have. They need to be warned and I consider it my job to be the person to do the warning. Pharmacists have a duty to warn. Lawsuits will be lost because there has been harm done and you did not warn.

I learned a lesson when I refused to refill a lorazepam prescription. He was a young Navy man in his late twenties. The prescription had refills, but it had only been a few days.

He screamed at me, "That's my seizure medicine. I have to have it."

Then his mother jumped me. "He could die. You can't refuse him. Where is Dave. I want to talk with Dave." Dave had quit the store months earlier.

Withdrawal from benzodiazepines can be life threatening. This sailor's entreaties just blew over my head. Seizures are a possible withdrawal symptom.

This kid did have a seizure and his burly father came in and threatened me with his own kind of seizure called an *ass-whooping*. Then the doctor called and dared to jump me.

"You're the one who wrote the instructions, Doctor. You wrote four tablets a day and he had to be taking twenty a day."

"You should have called me."

"No, Doctor. You wrote the prescription. I should not have called you."

Remember that drug seekers can be desperate and desperation can make for a risky situation.

Talk with them. Underneath the hunger for the drug that has their full attention is a human being. Keep the ability to refuse and still show some sensitivity. It doesn't hurt to smile. You can offer to contact the doctor.

Then there are the clear-eyed buyers. You are just the wholesaler. They are selling it. They get the prescriptions from *prescription mill doctors*. They pay cash.

Euthymic Ideation

Rule Number Four
Stop with the Medicalese

They have no idea what you are talking about, but they smile. They nod. They call you *Doctor* and they thank you very much. When you ask if they have any questions, they say, "No, I'm good." They just want to get away from you. They leave wondering what the hell is wrong with them.

It is not at all helpful to use the words you learned to love when you were suffering from an acute case of *Doctor Syndrome*.

I was an absolute jerk abut this. I loved the sound of multi-syllable words. I couldn't stop doing it. I was young. I was like a robot pretending that I was a scientist because I could speak to patients in un-decipherable *medicalese*.

Yes, I was that young. Twenty-three when I got my first license in Ohio. I laughed at a patient who thought that *hypertension* meant that he was supposed to be *nervous*. I had typed *hypertension* on hundreds of labels. The doctor had written it on the prescription (his own supercilious affectation) so I typed it on the label.

Hypertension. This was the age of the stroke. The best drugs were not very good. A middle aged man confessed, "I'm not nervous. I don't need it."

It doesn't matter what you tell them if they have no idea what you are telling them.

You read about it in a clinical article so you use what you learned. You tell a young mother, "Apparently, you child is suffering from *eosinophilic esophagitis and not gastro esophageal reflux disease"*. You think how cool this is and wonder why the mother is crying.

"Maam, I.. ah.. Can we go over the medicines... Ah."

Come on. You are an idiot. A well-educated mother might just put up her hand. "What the hell are you talking about?" The high school graduate with even average self-esteem will run away. You have ruined it and the patient just may suffer. You will never get this mother back. After the crying, she will nod and smile. You have shamed her.

Try cleaning your most-respected-practitioner-to-patient vocabulary of some of these: *Edema. Pruritis. Vertigo. Urticaria. Euthymic Ideation.* Unless you are hopeless, your patients will be much better off if they know what you are talking about.

Carmen Miranda was not her real name, but it will do. She was a naval officer's wife. Carmen dressed up just to go to the pharmacy. A dress, heels jewelry, the works. Carmen was a college educated beauty who refused to go to seed.

The prescription was for meclizine. When I got to the instructions, I typed exactly what the doctor had written.

Carmen was a very high maintenance patient. Her husband had been a flag officer, an admiral. Oak Harbor is the home of Whidbey Island Naval Air Station. Carmen had been the wife of a *rock star.*

One of the percs she expected being the wife of an admiral was a few minutes of face time with the pharmacist. I was able to manage only 30 seconds.

"Carmen, as usual, *belleza,* you are gorgeous today."

"*Gracias, Jim.* You are too kind. Yo soy solo una dama de edad."

"If you just said that you are a beautiful woman, I agree."

I went and made a judgment error on her Rx label. I depended on the auxiliary stickers to do the counseling. Had I counseled her, I would have used the words *ringing in the ears.* But, I didn't. I typed the word *tinnitus.* It was easier. I was busy. I had given Carmen her fanny pat. That should be enough.

I continued to work and happened to look up and see that Carmen was being comforted by her friend Elizabeth, another older naval officer's wife who dressed well to go shopping.

Elizabeth was talking with Carmen, but the admiral's wife was inconsolable.

I was frozen. I just stood there and watched the scene like the other 5 people gawking. What got me to move was a teen age girl. She was dressed in the Seattle grunge uniform of the time. She started to laugh and pointed at Carmen.

The older woman could not stop crying. Her body was in spasms. When I got out there, she looks at me and wails, I have *teeneetus*. Am I going to die, *Jeem*? Am I dying?"

Her friend said, "Her doctor said nothing about this disease. Is she really dying?"

"No, she has ringing in her ears. *Tinnitus*."

"*Euthymic Ideation*, by the way, means *normal thinking*.

Rule Number Five
Leave Me Alone

You tell me why you cannot enjoy peace and quiet like everybody else while you are looking over the red and yellow sweet peppers at the Cajun King I.G.A. Supermarket? You had to put up with the Medicaid link being down for eight hours. The District Manager called and said, "Get it fixed." He hung up on you. He was talking about the pharmacy being over budget on technician hours. The store manager wanted to know why pharmacy staff members are not on the bathroom cleaning schedule.

All you want is to go home, prepare a nice Thai stir fry with brown rice and enjoy a glass of white wine. It is your one chance this week to have a nice meal with your family.

The red and yellow peppers are on sale. You do well, but you are still frugal. You have not always had this kind of money. 89 cents for red and yellow peppers thrills you.

This is so much fun. Just lazying around in the produce. You are usually in a hurry. You pick out the best Vidalia onions, a handful of fresh snow peas, some nice bok chow, a few fresh shitake mushrooms. You take a big root of fresh ginger, some Thai spices and some giant shrimp. You decide that you will do this more often. It just feels so good.

When was the last time you saw a lawyer interrupt a leisurely session at the bell peppers to answer a legal question? You laugh so hard I worry you will hurt yourself.

"Ah...hello, Mrs. Green." *Oh gawd, no. Not her.*

"Oh just call me Phyllis. I call you Becky, so we are friends, don't you think?"

You give her a frozen smile. How to get away?

"I am so glad to see you Becky. My neighbor Sally, you know Sally Frost. She has an itch in her..well her..private areas. What should she do, Becky?"

The smile on your face is so frozen that it is almost painful. Your neck heats up. *Don't touch me.* You step back. "I am in a hurry," Mrs. Green." *You are not in a hurry.* "I have to be somewhere?" *There is no place to be other than your own kitchen with the cutting board and your good knife.*

"You know Sally. The cute Cape Code on Wetherly."

"I am sorry. I have an appointment." You rush to the check stand, like an idiot who just allowed a woman you know only professionally to ruin a enjoyable occasion. Plus, you forgot the brown rice. You settle on Minute Rice from 7-11.

I have been called at home on Thanksgiving Day. A lawyer's wife bothered me so often that I hit her husband up for fee legal advice *and got it.*

I have been asked for medical advice in the movie line, at the gas station and too many times at the super market. One guy took my recommendation and wanted me to go to the OTC drug section with him. There have been times without a *thank you.*

What a bargain the pharmacist's fee is.

It is not too late for you. A doctor will sniff and say, "Call my office tomorrow morning. There should be something right after lunch." The attorney just sniffs.

Here is something, "Please have Sally Frost stop at the pharmacy tomorrow morning". I will find time for *her.*"

It is your life. It is time to be a zealot about it. You give free advice on the job. You do not have to allow them to disturb your real life. Especially when you are having fun.

Rule Number Six
Listen, Really Listen

Multi-tasking seems to be a necessary talent in modern America. Children learn when they are given their first cell phone. They text at the breakfast table. They talk to a friend while they are searching the World Wide Web for information for a report. This is what we do in America.

A pharmacist who cannot multi-task may have a hard time in a modern drug store. You have to learn how to compartmentalize your attention. You have to set aside the need to apologize to your spouse after acting like a knave the night before. You had expected fish and chips at 10:00 PM. She had meatloaf and you refused to eat it.

At work, the normal job of a pharmacist demands that you do multiple tasks all at the same time.

You know exactly what I am talking about. You check prescriptions with the phone glued to your ear at the same time that you are watching a senior citizen on Aisle 9 shoplifting a bottle of Centrum Silver. The technician reminds you that she has to leave for a dental appointment in ten minutes. A customer out front yells, "Hey, you! Where's the beer on sale?"

"This is not a good time," you tell the tech. "Use your head when you schedule appointments."

Most pharmacists have well developed talents in multitasking. The ability to partition your attention is one of the attributes that make you successful at this demanding job. You may be able to keep multiple balls in the air at one time but you have most likely become a lousy listener. When you are having a conversation with your husband, about things that are vital to *him*, you have to get out of *computer time* and get back into *human time* or there will eventually be problems.

It is your spouse's primal complaint. "You don't listen to me!"

Human beings do not usually give immediate feedback. They hesitate. They smile. They laugh. They look at you and then at their feet. They may be embarrassed. You need to listen intently or they will give up.

This is not a time for you to impatiently say, "What else?" You don't need to be studying your fingernails or tapping your toe. You can squash her courage simply by not looking at her. A yawn is like ice water. She will know that you are not with her and *this may be exactly what she wants.* She never really wanted to tell you about the little lump she found in her breast. It is too scary. It will go away.

I still, as of an hour ago, have to discipline myself to act like a human being. Especially at home with Victoria. I get going so fast that I just can't stop dancing. There have been days when I have been at home decompressing and I get a sick feeling because I never really answered a patient's question. I let myself get distracted by the different tasks that I had to do and simply forgot to return to his needs. I clearly recall the forlorn look on his weathered face when I walked away. I had to get a doctor call and never returned.

Every circus needs a clown and I am that man. Don't be so smug. You are no David Letterman yourself. I've missed plenty of chances to make a difference. This is a flaw in my game. I am having to reset often. The problem is that I am not that thoughtful. I know that I am always harping on the importance of being fully human and available to patients, but I am flawed. I still have to struggle to listen to my wife. My brain wants the excitement of the *trailers* all the time.

Some of the most satisfying events in my professional career have taken place when I have been lucky enough to catch myself listening. I say *lucky* because I think it happened by accident. My attention is usually fractured all over the place.

I am not consistent. At work, I am a frenetic, driven, multi-tasking, determined-to-stay-on-mission, get-out-of-my-way pharmacist. Who doesn't look back often enough?

I *know* the value in listening. That is in my favor. The patient benefits when I look in her eyes and listen to her concerns. She is often living with wives' tales. My daughter-in-law from The Balkans still believes that temperature will make my grandson sick. This is Galveston and she still bundles him up. Ivana trusts me on all things
Medical except that one.

When a patient (usually a woman) thanks you for listening with a weepy smile, there is nothing like it. The doctor did not have time. The nurse swept her away, but, down at the bottom of the funnel, is the pharmacist.

The next time you notice that you aren't a very good listener, you are there. None of us is perfect. Someone once told me, "If you want to be interesting, be interested.

Rule Number Seven
Don't Be an Idiot, You are Not a Hero

Other than a particular Mormon pharmacist in Santa Rita, California, none of us are hero material. The Mormon pharmacist saw the gun, grabbed his own gun and **Blammo** shot the robber right between the eyes. This robber was a very mean man. The police said that he had killed two pharmacists as he worked his way down the coast from Seattle.

I did exactly what he wanted, but not fast enough. He pistol whipped me. The two dead pharmacists had resisted. Did they see too many movies? Was Dirty Harry on their minds. The Mormon never even said *Make my day!* He just shot him dead. It splattered blood and brains all over the Whitman's Samplers. I suppose a technician cleaned up the mess.

I have been robbed at gunpoint three times. All were in the 1970s in California. A pharmacist in Concord, California was shot in the head for two cartons of cigarettes. He was a friend.

It is vital that you know exactly what you will do if you ever have a gun pointed at you. Hysteria could get you pistol-whipped. Defiance could get you dead. If you have not thought out what you will do in a robbery, you might just freeze.

This is what happened to me. *Fight or flight* chemicals surged into my veins. My heart pounded. My mouth was immediately parched. My extremities got ice cold as my body routed the blood to the vital organs. My head buzzed like a thousand bees. I could barely hear the orders that were shouted at me. You have to be able to hear so you can obey. I recommend that you obey.

The first robber was a long-haired, unshaven scruffy kid. His gun hand trembled and that worried me a lot. This guy could pull the trigger by accident. I had not thought about this ever happening and I was frozen. Somehow, we got through it, him and me.

He wanted drugs. No money. This was pre DEA. He got 1000s of pentobarbital, secobarbital, meprobamate, benzos and every tablet of narcotics that we had. Months later, I saw him in court. Clean-cut in a suit. His mother sitting weeping behind him and his lawyer.

That kid caused me to think about being robbed. The second robber was the guy the Mormon shot. I had decided after that first robbery that I would do whatever a robber wanted me to do. I was just too slow. He hit me with his pistol, above my left ear.

This guy wanted both money and drugs. He got everything he wanted. I filled the grocery bags for him. Then he tied us all up. I guess so he could finish up at his leisure. When he ran out of rope, he looked at the teenage stock boy and said, "I guess I'll just shoot this guy."

Looking back, I can see that the Mormon pharmacist just got lucky. I do think that the robber would have shot Eric in the head, where he had the gun, if a fast-thinking clerk had not said, "Use Scotch Tape".

By the third time faced with a gun, I knew exactly what I would do. I was prepared. Two henchmen stood at the front registers to make sure that business went on as usual. This robber was a professional, very calm and polite, almost collegial with me, like we were fraternity bothers.

I considered myself to be his partner. I asked him what he wanted and it was only money. The store was a collection point for cash payments to Pacific Gas & Electric. There was a lot of money. Close to $30,000.00 in cash. Then he took watches and diamond rings. When he took the cash from my wallet, he said, "I don't want your ID or credit cards. You'll have them cancelled in ten minutes." What will you do? Let a robbery be an inconvenience, a speed bump in your career. Don't *wing* this one. Be prepared. Give the robber everything he wants. Show him the hiding place the boss doesn't want revealed. No heroes. The pharmacist in Concord resisted two teens who wanted cigarettes. The .22 caliber bullet glanced off his skull. Pure luck!

Rule Number Eight
Write It Down

Document! Document more and then document even more. You can't even remember the details of this morning, you are so busy. How can you be expected to recall what happened a week ago? A year ago when that non-pharmacist store manager caught you in a corner and demanded a kiss? You quit that dream job because of him and now the E.E.O.C. wants the details. This is not a lightweight suggestion.

Five years ago, a state board investigator wanted to know the details of a dispensing error I made during a hectic morning. The patient had it in for me. She declared that I was incompetent. She accused me of unprofessional behavior. What really pissed her off was that I took a couple trips to the sink area to take bites of my peanut butter sandwich while I worked on her Rx.

My error was filling the Rx for amoxicillin when it was

written for dicloxacillin. *I am the one who caught the error.* I was in the habit of looking over all of the Rx I have filled at the end of the day. There it was. I saw the error. *I called the patient and offered to deliver the correct Rx after the store closed.*

She refused my offer and did not pick up the corrected Rx for 4 days. Prominent in her letter to the Vermont Board of Pharmacy was that I dared to be eating a peanut butter sandwich while I worked on her prescription.

This was a small village pharmacy. There was no technician and there was no cashier. The volume was rarely over 100 Rxs and the pharmacist did everything.

I documented everything about this prescription. The investigator made noises like Ah Ha and Ummm as I told him my side of the story. He was all smiles and gave me reason to feel good, but he still had to take it to the board. I followed up with a complete written account. I was very comfortable with this part of the process because I had all of the details and I had the details because I had documentation.

If there is no refill on an indomethacin prescription and the pharmacy manager says, "That doctor always okays that drug. Go ahead and fill it. I'll get the okay later." If you choose to fill the Rx rather than cause a problem between you and the pharmacy manager, you better document what happened to cover your ass.

You have legal rights about the hours you work and the pay you receive. Never work off the clock. Period! Document every single minute you work on your calendar and keep the calendars forever. Research the laws in your state. Some states mandate time and a half over eight hours in a day or more than forty hours in a week. In some states, the premium is mandated only after forty hours.

Inappropriate behavior of any manager or colleague is never acceptable. Document the first incident. Then, you must inform the offender that the behavior is unwanted. If you can't say it face to face with a witness, then send a letter Certified Mail. If the unwanted behavior continues, document each and every instance.

If you wonder if a bottle of methadone is missing. Document. I worked with a non-pharmacist store manager who opened the pharmacy because he claimed that he heard the water running. The two pharmacists suspected that he was lying. We informed our DM, but were reminded that his grandfather was the founder of our billion dollar company. All we could do was document. Any error made by anyone. Document.

The job interview is where the company will make promises to you. They will almost always live up to the promises they make about wages, holidays, sick pay and other solid benefits. They will almost always fail to live up to the promises they make about your vacations.

I don't know if there is any way you can get the companies to come through on vacation promises. It can't hurt, however, if you have detailed documentation of your discussions during the interview as well as the name of the person making the promises.

Rule Number Nine
Show Some Respect
Colleagues Not Competitors

What it is with these pharmacists who act like we should hate each other? The ones who make us wait five minutes for a transfer and then give it grudgingly. We are not enemies. We are colleagues!

I am ashamed that I once put the growth of a business I managed first and watched a competitor go belly up. The owner was in jail for midnight trips to steal plants from a nursery. His wife was not experienced and the only pharmacist she could get could not get a job anywhere else. He was barely competent. I smelled blood. She had to sell. I will forever not be proud of that.

Recently, I called a chain that is not Three Pee Ex and

asked for a transfer. I had the vial and all I needed was for the pharmacist to verify the information and to give me the insurance information. The patient did not have his card and said, "My insurance is *on the computer at Three Pee Ex.*"

The pharmacist answered the telephone. I gave him the Rx number. "Why'd you call on the doctor's line?"

"Because you don't have a pharmacist's line."

"I'll be right back" and I was on hold, listening to soulful music for ten full minutes." I held the phone between my left ear and left shoulder. I have done that for so long that it is no wonder that my neck hurts. After we had the transfer legal, I asked for the insurance information.

"I'm not giving you that."

"What?" *Are we back in the 1970s?*

"We took the time and effort to get that information. You can get it that way."

"Are you serious, man?"

"Very serious." Slam went the phone.

This was 30 years beyond the low point of pharmacy's era of cutthroat competition. What's wrong with us? With 80% of Rxs being 3rd party, there is no price battle anymore. Sam's Club and Harriet's Drug Store collect the same copay. You get patients for other reasons. You get and keep patients for convenience, delivery, house charge accounts, D.M.E, good old fashioned service. Personality. Independents have marvelous opportunities in the 21st Century.

Pharmacy was a profession in which the practitioners cooperated in almost all affairs until the mid 1960s. It was a 40% profit business. It was not honorable to cut prices. When you wrote a *copia,* you would write your retail price in code (PHARMOCIST) on the face of the copy. You could count on it that the next pharmacy to fill the Rx would charge the same price. Our business behavior was principled and praiseworthy.

Then came the deep-discounters. The goose was cooked. All of a sudden, pharmacists started being nasty to each other. Prices were slashed. Price became the entire game. Forty percenters adjusted or they died. They eliminated inventory. They cut the lights, the heat. Lunch counters were closed. The golden age of pharmacy was dying. In Ashtabula, Ohio the six pharmacies in a three block area in 1959 would not all make it.. In 1965, there were only three. Today, there is one.

Into the 1990s, pharmacists smiled, but were likely to have a dagger to stick in each other's backs. There is some of that left, but most of us know that we are family. There are tensions, but the general sentiment is respect. I never hesitate to get the phone when there is a transfer. Why be a jerk because our companies are fighting it out. I talk to other pharmacists. I find out how they are doing.

The pirates are rare these days, but you will know one when you see her. Mutual respect. Even admiration. They feel good and why not feel good? Pull together.

Rule Number Ten
Do Not Tolerate Abuse or Harassment

This story is indelicate and I am very uncomfortable telling it. You are going to say, "What is he talking about? Calling me an idiot? He was an idiot if there ever was one."

I was, indeed, an idiot. I'll tell you the story, but first.

Abuse from patients is something you need to walk away

from. No explanations. Just turn your back and go. You are a pharmacist for goodness sake. A highly trained and educated medical professional. You know your job. You are well paid. I don't care what your company says. You do not have to stand there and let any clown work you over. That includes the college professor or the out-of-work Medicaid recipient. An "F" bomb is abuse no matter who says it.

They can really get started when they have no refills.

"Sir, the refills end at 6 months on Androgel."

"Just fill the goddam prescription."

"Sir, I can't. It is the law."

"Piss on the law, girly! The doctor is the boss." He slurs and stabs his finger at you. "What the hell are you saying? It says right here, on my package, one refill remaining." He flexes his shoulder muscles nervously and you notice that his eyes are a yellowy red. He starts to rattle the door into the pharmacy. Can transdermal testosterone cause *roid rage?*

I think that girly needed to walk away at *goddam*. She does not have to let anyone abuse her. It will just escalate. The patient doesn't seem to understand that the issue is non-negotiable. There are men who are used to pushing women around and, unfortunately, there are plenty of female pharmacists who are used to being abused by men. It can be a "Daddy Thing".

I have watched female pharmacists just stand there and take it. Frozen while an aggressive male verbally abuses her. She is almost unable to assert her right to some little bit of dignity by just walking away.

I know. The non-pharmacist store manager will always blame you if there is a complaint. This time, tell him that you are not putting up with anything from anyone, including him. If the manager won't protect you, his most important employee, well.. go take a 30 minute break. Who cares if there are six people waiting. Don't go back to work until he apologizes to you. That is *brinksmanship*. It works.

Set your boundaries early. Every adult has limits or should have them. Assert yourself vigorously and you will never have to again.

Sexual harassment no longer works only in one direction. It runs both ways. You female pharmacy managers who have been sleeping with that handsome intern, watch out.

Tommy Boy could have your ass when your company is ordered to pay damages and this kid is able to pay off his student loans with one check. You can't win.

Running his fingers through your hair is not normal management behavior. Put a stop to it right now. If you laugh off harassment, it will only be exponentially more difficult to bring to an end. In a letter sent by Certified Mail, just mention the words *Employment Equal Commission Opportunity* all together, in the right order, in one sentence and watch them jump.

Boys, you may think that a pat on the fanny from an attractive female manager is sort of cool, a compliment, but it is harassment if it is unwanted and you can get that company to pay for your ride for a long time if it continues.

You are a pharmacist, the most important employee that a drug store company employs. Without you, they can't even call it a drug store.

My indelicate story. In the men's room, the store manager said from the stall, "Plagakis, I'm in here hatching a new pharmacist." He chortled. He thought he was so funny.

I walked away. It emboldened him. I am ashamed. I was young, still an idiot.

Rule Number Eleven
Counsel With Care

"That can't be right", the older woman complains, "My doctor would never give me anything dangerous."

"Maam, I am sorry, but just about everything we have back here is a poison."

"What? Poison? I don't want a poison. My doctor would never make me take a poison." She is irate. Her face is red.

"Maam, listen to me. It is dose and frequency. Most of the drugs we dispense can kill you if you take enough of them often enough."

"And that is supposed to make me happy?" She laughed.

"We just poison you a tiny little bit." I laughed with her. "In this case, it is a half milligram. That's an amount so small that you can't see it with the naked eye."

Her eyebrows raised. "Okay," she sighed. "I get it. I'll take it like you say I should."

There is a lot that can go wrong when you are counseling and it is not the facts. Side effects, dosage, contraindications, special instructions are what they are. Black box warnings make you pay more attention. You give the patient more information and you make sure that they get it. You can even refuse to dispense. The days of blindly giving out drugs is long gone. Harming the patient is not unheard of. You can tell the prescriber if she is wrong. Doctors are not infallible, but neither are you.

I had a perfect storm of counseling-gone-wrong a few decades ago. My day was ruined. The patient was dumbstruck and her family was mortified. The doctor made an inappropriate, even stupid choice, but I was too helpful.

I gave too much information. Remember, this was in the early 1970s. Doctors were not accustomed to pharmacists taking the lead on assuring satisfactory outcomes. In the 21st Century, there is not too much information.

Use your head. Consider that many of your patients are just not sophisticated enough to understand what you are telling them. You are very smart. You proved that by getting through pharmacy school. Many of your patients live their lives in intellectual pits. If you have a thing about being humble, get over it.

Humility is a desirable attribute in your real life. There is value and spiritual growth in modesty but, in your professional life stay grounded. You are educated.

Your schooling was six years. Dummies need not apply. You are smart and capable. You know a lot, but too often your patients may not know what you are talking about.

Recently, I went over an albuterol HFA MDI with a young mother. I moderated my counseling so she could understand what she had to do. She nodded and smiled and made all of the right verbal noises. Ah ha. umm, Oh and okay. I had a prickly sensation in my neck when I was done so I asked her to repeat the basics to me.

She could not even start. "Um, I, well she puffs it."

"You have no idea what to do with this thing, do you."

Her face was scarlet. "It's complicated. My mom knows though. She can help me."   This woman was ready to leave the store without a clue. I turned the box over and we went over the instructions slowly and carefully. I asked her to read the instructions once more, out loud, to me.

She started to tear up. Her lower lip trembled. Clearly, she was about to cry. She could not read.

I did visual counseling. The MDI came out of the box. I fit the spacer on. I shook it. I actually gave the child a dose.

Look at their eyes. Most patients can get *four times a day on an empty stomach*. What about Advair or Exubera. Has anyone even dispensed Exubera?

Remember that you are most likely smarter than they are. You certainly have more education than most of your patients. The prescriber probably told them very little. You are at the bottom of the funnel. It's up to you.

Back to that dreadful story. I was counseling a middle-aged woman on her Leukeran therapy. This was around 1987. I rarely dispensed this drug so I brought out the package insert. We went over what I thought was important. She listened attentively. I was confident that she understood. Then she asked me a question. Her eyes darkened.

"What is this for, Jim?"

"Well.. it's.. It's chemotherapy, Joyce."

Her eyes widened, "He told me it is for my blood."
The husband said, "Is this cancer?"
The doctor was old. He was not an oncologist. He did not want to burden Joyce's family.

Rule Number Twelve
You Earned It, Take It

When you are sitting face to face with a pharmacist recruiter, that is the time to record all the promises that this company is making. Not the least important is the generous vacation benefit.

Every other benefit falls right into line. A 401k with 5% company matching. Medical, vision and dental from day one. You get a BMW sedan and the lease is paid as long as you continue to work for the company. You kept a straight face.

Much better than expected. The vacation sealed the deal. More weeks than any other company promised. A hand shake and you were onboard.

Stop! The hand shake is not good enough. You find that out after your have put in a year. It is time for your first vacation and you really need it. Twelve hour shifts have been piling up. You worked with a miserable head cold. The relief pharmacist was pulled during the profit crunch. That meant that you had to increase your speed. The Pharmacist In Charge was in flight or fight. He jumped and twitched and whined about it, but did absolutely nothing.

He was a *corporate man*. He walked with a pronounced limp after four decades. His shoulders were bent. His *head forward posture* from working over a counter was causing considerable pain. His shuffling gait and periodic groans worried you. He was younger than your father. You had been observing the PIC and promised yourself that you would never get like him. Your two week vacation would heal all.

They promised you the generous vacations, but no one told you how hard it would be to get the time off. You are a new hire. The idea of getting to take your vacation during the summer months when the kids are out of school makes the company veterans laugh so hard they hurt themselves.

The vacation coordinator at the head office suggests that you take a winter vacation. "San Diego is really nice no matter what month you go. The San Diego Zoo. Have you heard of it?"

"Of course, I have." You try to chill her out by being silent, but she is good at this.

"Your kids will love Sea World and there are bargains to be had at the best Mission Bay resorts. The beaches are to die for."

Hmm. Not a bad idea. "Put me down for December. Two weeks on both ends of Christmas."

This is a helluva good idea. Peggy will love getting sun in December. A little Christmas tree in the hotel suite, all decorated. Breakfast from room service for Christmas morning. *Huevos Ranchero.* This is San Diego not Milwaukee. Oatmeal and a bagel will not do.

The kids will be overjoyed to hear this. A swim in the pool before opening presents. This is a terrific idea. Julie may be 15, but she is not wearing.... Why is this woman laughing?

"Oh, Tony, I am so sorry". She is still chuckling.

"My name is Anthony. No one has called me Tony for 20 years."

"I'm sorry, Anthony, but December is out. Company policy. No vacations in December."

A long silence. Is she asking me to settle? I was told that they would offer me money to just forget all about the vacation.

They encourage taking your vacation in short sips. A four day weekend. It is good for them and lousy for you. Nobody can decompress from this job in four days.

"I don't care about a policy that says no vacations in December. I am going home today and I am lighting up Orbitz. I will have reservations for flights and a hotel in San Diego at Christmas before I eat my dinner."

"But..But.. you can't do that. Vacations are not allowed in December. It's company policy. It's just not done."

"Maybe I should look elsewhere for work. Maybe I made a mistake agreeing to work for CostLess Drug Stores. Perhaps I need to take a second look at the promises CostLess made. I have them documented."

"Uh.. Anthony.. uh. I will have to talk with Mister Jones to see if...."

"No *Ifs.* It is a done deal. Christmas in San Diego this year for me and my family. You have six months to find my replacement. Goodbye, Harriet."

Now, how good did that feel? It is your vacation. You earned it. They made promises. Take it when it works for you and your family. An inflexible company may not be where you want to spend your entire career.

Rule Number Thirteen
They are Dying

Very few people are willing to talk, really talk, to a terminal patient about their situation. "How does it feel to be dying?" A question this forward must be asked gently. There must be a relationship with the patient.

"Awful," she will look right at you. "Then it can be peaceful too."

Pharmacists rarely ask the patient if he is frightened You can talk about the pain and the nausea, but you hesitate asking where she got her wig. Is her will in order? Has her family listened to how she wants her memorial arranged?

There are loved ones who can be mean, "You know that I won't talk about that. It is macabre. Why do you have to be so damned depressing all the time?"

If you Can't Talk With Them, At Least Listen

It is uncomfortable, to talk to a terminal patient. The patient avoids the subject because she doesn't want to ruin anyone's day. We admire their courage.

"That Casey Quinn is amazing. Always smiling. She never complains."

Casey is a true champion of the game of life and of dying. If there is a one hour support group, they may get to unload on people who understand.

"I might as well be dead already. I hate my family. They have these asshole smiles all the time, 'Dad, you look good this morning. Are you coming to my game?'"

"I hate them sometimes. I feel like shit, but if I want to talk about it, they close down and disappear."

In September, 1964, my Ohio license was so fresh the ink was wet. I was at a cocktail party in a garden. People with highballs in their hands were moving like the tide from room to room. There was soft piano and cello music from the veranda. I had my eye on a blonde Finnish girl. I had liked her in high school, but I was too shy. I was a pharmacist now. Maybe I had a chance.

A woman walked up to me. She was skeletal. Her flat blonde hair looked uncombed. Her smile was luminous. She leaned toward me and kissed me on the cheek. I knew who she was, but I was a coward.

"Do I know you?" I asked.

"Of course you do, Jimmy. I am Millie Blake." Her smiled warmed me. I could have fallen in love with that smile all over again, very easily.

I stammered, "I didn't.."

"...Recognize me?"

I cringed with embarrassment. We talked for awhile and

then she took my hand and put it on her head. I felt the tumors. I pulled my hand away quickly. I must have frowned. I stepped back. I was so young.

A shadow darkened Millie's face. "It's just my cancer, Jimmy." She studied me. "It's not *catchy*. Her eyes were questioning. "You're still my Jimmy and now you are a *pharmacist*. Why are you afraid?"

"I.. I heard you had cancer."

"Not *had*, Jimmy. I *have* cancer." She managed a smile, stepped very close, looked up and kissed me wetly, full on the lips. It was a long kiss and I felt her tears on my cheek. That is the way it had always been with us. Stolen kisses, embraces in the hall. She was older with two children and a high school principal husband. We respected that. I love her memory to this day.

I never saw Millie again and I have never been able to forgive myself for being such an insensitive lout. I actually washed my hands. I felt contaminated from touching her head.

I learned a difficult lesson very early. *They are still alive.* I became at ease with terminal patients. I have visited them in their homes. I have asked the questions that they want to answer, *Are you afraid? Does it hurt? Heaven?*

They tell me how embarrassed they are around friends and family members. They don't mean to make people so uncomfortable. We can be such pigs. It is almost like: *Will you please hurry and die.*

The family of the terminal patients can get the pariah treatment too. They need to talk and I do not care about how many minutes you are away from "The Prescription Mill" You need to listen. The floodgates will open. You become a very special person. They will love you. Real communication can do that.

Their doctor will never just listen. The nurse talks about the game Friday night. You are at the bottom of the funnel. Everything falls to the pharmacist. You can make a difference.

Rule Number Fourteen
Like it or Not,
You are in Business

Get over it. You are in a business whether you like it or not. Dentists know that they are running businesses. So do surgeons. Some of you want to think that you are not tainted by the *business* of pharmacy. You declined the job of Pharmacy Manager more than once. It is not *professional.* Too many reports. Too many details. Too many budgets. Too many supervisory particulars. You don't want to hire people and you certainly don't want to fire anyone. You are a *professional.*

You play golf with a lawyer friend and you feel squirmy that he will find out that a big part of your job is to order pharmaceuticals and to keep the return on investment satisfying to the investors. That's the game isn't it? I suppose you think that the attorney is not in business. How about $250.00 per hour billing in $42.00 ten minute increments?

You want to make sure that he knows about the immunization services that you have been trained to provide. You administer flu shots, tetanus, pneumonia, Gardisil, zoster. The works. Now, that is *professional.* The Pharmacy Manager is practiced in Diabetes MTM. Every body does diabetes.

Two months ago, you became the last word on psychopharmaceutical MTM. That is rare. So rare that there have been no clients.

You do not like it that the manager has all of her personal

things all over the desk in the counseling office. She has family pictures on the desk and those are *her* certificates and licenses on the wall. Why should she do her reports and other *business* at that desk? It is the *counseling* office. You could do a much better job than she does.

Sorry. She is the boss. She has worked just as hard as you have. She is just as *professional* as you and she probably actually enjoys her job a lot more than you do.

When you sell a commodity, you are in business. Don't give me that look, you hospital high and mighties. What do you charge for an ampoule of atropine? Don't think for a second that the Director of Pharmacy gets away from dollars and cents when he attends meetings.

During the first couple years of practice, I thought I would escape the stain of *business*. I tried to look the part of a scientist. My day to day costume was my white Mister Barco jacket and a button down blue shirt and a regimental tie. One day, I was mixing up some Dakin's Solution. I dribbled some on my shirt and tie. A five dollar Arrow shirt and a new tie. I avoided compounding Dakin's after that.

Mister Professional. The one talent that best defines a pharmacist is still compounding and I avoided compounding that could get my clothes and carefully drawn image dirty

I eventually figured out that pharmacy is every bit a business as it is a *profession* and surrendered. My first managerial job came just 5 years after I had first stepped behind the counter as a registered man. I was good at it and it did not detract from my professionalism.   Let's define Professional. *Engaged in an occupation for pay and not a hobby.* Sounds like a business to me. *An occupation requiring special education.* You are always going to be a *professional.* It doesn't matter how many reports you complete, how many budgets you reconcile with the schedule, how many time you empty the trash and wash the compounding glassware, nobody can say that what you do are *unprofessional tasks.* Every chore that a pharmacist engages in is a *professional* task.

A technician asked me recently, "Jim, why do you empty the trash every day when you get to work?"

I told her, "Because it is full every day when I get to work."

Doctoring and dentisting are businesses too. Profit is the reason for the season for every MDs, DOs and DDS. It's a living. A minority of the residents at UTMB talk about getting into a specialty where the hours are nine to five Monday through Friday and the money is big. Dermatology and Plastic Surgery come to mind. They look at me funny when I tell them that they should have been a dentist if money was what they wanted.

Pharmacy is still a *profession* first, but the business aspect can dominate if you let it. Look into a job at a Poison Control Center or a call center at any PBM. There are other choices in which you can avoid the commerce of the *profession*. Get off your duffs and look.

Blue Cross Blue Shield of Michigan

Medicare PLUS Blue PPO℠

(1) Enrollee Name
VALUED CUSTOMER

Plan H9572_001

(2) Enrollee ID
XYO888888888

Issuer (80840)
9101003777

RxBIN 610014
RxPCN MEDDPRIME
RxGrp BCBSMAN **(3)**

Group Number
50802

MA|PPO
MEDICARE ADVANTAGE

MedicareRx
Prescription Drug Coverage

Members: bcbsm.com/medicare   Providers: bcbsm.com/ma

Blue Cross Blue Shield of Michigan
A nonprofit corporation and independent licensee
of the Blue Cross and Blue Shield Association

Michigan health providers bill:
BCBSM - P.O. Box 440
Southfield, MI 48037-0440

Mail DME/P&O claims to:
P.O. Box 81700
Rochester, MI 48308-1700

Mail pharmacy claims to:
P.O. Box 14712
Lexington, KY 40512

Medco
Pharmacy Benefits Administrator

Customer Service: 877-241-2583
TTY/TDD: 800-579-0235 **(4)**
Misuse may result in prosecution.
If you suspect fraud, call   888-650-8136
To locate participating
providers outside of Michigan: 800-810-2583
Provider Inquiries: 800-676-BLUE
Facility Prenotification   800-572-3413
DME & P&O Providers: 888-828-7858
Rx Prior Authorizations: 800-437-3803
Pharmacists/Rx Claims: 800-922-1557
Use of this card is subject to terms of applicable
contracts, conditions and user agreements. BCBSM
assumes no financial risk on ASC claims. Medicare
charge limitations may apply. Out-of-state providers:
file with your local plan.

Rule Number Eighteen
Not Your Problem

    This is definitely heretical thinking. How dare Plagakis tell you that insurance problems are *not your problem*? Everyone knows that dealing with the PBMs is just part of a pharmacist's job. PAs. Rejections. Days supply. Vacation overrides. Especially when a mouthy patient calls *you* over to the counter because she doesn't like what the technician just told her.

    "I have insurance. I got prescriptions for my husband

last month and my copay is $5.00. I never pay more than $5.00. You have to fix this. I have never paid more th...."

"I am sorry, Maam. The technicians are the experts on all insurances and copays. Please excuse me. Mrs. Martinez was here before you and I am going to counsel her."

The woman just stares at you.

"Privately, Maam. If you will please move away from the counter so Mrs. Martinez can get some space.

"Well, I never...."

Are you so institutionalized that all you can do is take a deep, defeated breath and get on the telephone to listen to Dvorak's New World Symphony while you wait and wait and wait? Do you really have to explain the issue to four different people? That is part of their Do-Not-Pay strategy.

What is so wrong about putting your patients who need counseling or OTC assistance first? Why not say, "Not my problem!"

What is wrong with telling the patient, "Call the number on the back of your card. Ask them why you do not have a $5.00 copay."

A few minutes later, "What does *cardholder ID invalid* mean?"

"It means that your husband does not currently have insurance with this company."

"My husband is a founding partner of Biggun, Doofus and Bees. Of course he has insurance."

This patient is new. You now know why she transferred prescriptions from both CVS and Kroger.

"Maam, you need to discuss this with the Benefits Coordinator at your husband's office."

"What does that mean?"

"It means that somebody screwed up because Mr. Doofus has no coverage. I can't help you further. Call the office. Mr. Doofus has no insurance."

"What?" It is a shriek.

Mrs. Martinez actually throws her shoulder into the fray and gets herself to the counter. She rolls her eyes.

"No insurance?" This woman actually has flecks of foam at the corners of her mouth.

"Calm down, Maam. There is an 800 number on the back of your card. You can...."

"..Don't you tell me to calm down. How dare you? Aren't you the smart one." She actually stamps her foot. I insist that you call my insurance company."

"No!" You move with Mrs. Martinez to the end of the counter.

Mrs. Doofus is still shouting at you. "Why won't pharmacists call anymore? The other pharmacists won't call. Pharmacists have always called my insurance."

The technician steps up. "I call insurance companies between   12:30 PM and 3:00 PM. I will put you on the list for tomorrow.

"Call right now."

"It's after 3:00 PM, Mrs. Doofus." The technician hands her the insurance card."

Mrs. Doofus leaps backward. "Why are you giving my card back? You haven't filled my husband's prescriptions yet. The copay is $5.00 for generic or brand name. I can get a 90 days supply for $5.00.

The technician smiles. "The 800 number is on the back of the card, Mrs. Doofus."

"Don't *Mrs. Doofus me,* you, you smarty-pants."

And that my friends is what *could* happen if pharmacists quit pandering to the PBMs. Pander, by the way, is **pimp,** just spelled in a different way.

The PBMs have a helluva deal. We have shown them that pharmacy will continue to allow them to take more. They are now testing us with reimbursements less than what we pay for a drug. We allow them to tighten their computers like Vegas tightens slot machines. But, hey, we own the casino. Are we all idiots? By all I refer to the owners and chain middle managers who sign the contracts.

The standard is that we kiss PBM ass. Are we going to stay stupid forever?

• Trust Me •

I'm a doctor

Rule Number Sixteen
You are not an M.D.

"Hey, you. You a druggist?"
"I'm a pharmacist, Sir."
"That's what I said. Druggist."
They look at you, grin, and say, *I trust you druggists more than I trust doctors. .* That means <u>You are free.</u> *You went to school as long as they did, didn't you?* Ah, stroke the pharmacist and it still means <u>You are free.</u> *You know as much as doctors do, don't you?* You know what is coming next and it still means <u>You are free.</u> *Will you look at this for me? It itches so bad that I got it all bloody scratching. Sometimes it gets all yellow and goopy.*

The next thing they want is for you to give them medical advice. You are free, remember. We still, after decades, have not figured this out. We provide a valuable service, saving the medical system millions every day and we don't get paid. Think about it.

Nobody else takes a look for free. Nobody else recommends an OTC, directs them to the doctor or sends them on a fast track to emergency. I had this guy with a terrible cut on his hand. It was a defensive wound from a knife fight over a woman. He asked me to bandage it..

"You better get to Emergency, man."
"I ain't got no money."
"Well you ain't gonna have no hand."

These people may actually be seriously ill and want you to diagnose their condition because they don't have any money or are desperately hoping that the pain really is just gas. They will want you to solve their problem with an OTC remedy. *You are just as good as a doctor.* Don't forget that, Druggist.

They take their clothes off. They try. Is it appropriate to show the pharmacist your big toe that is uglified from fungus?

They demand that you look in their child's throat. *He can't eat. He cries even when I give him marshmallows. His favorite is Diet coke and he won't even drink that.* If you tell this mother that she needs to get her kid to the doctor, will she just go down the street to the next drug store?

A heavy set woman, sweaty and growling, pulled open her blouse and lifted a hefty breast for me to be able to see under it. She had a moldy looking rash. The area had not seen the light of day for years. It was wet and rarely washed.

I trust I stepped back. It was years ago. I remember exactly what I said, however, because I never say anything else.

"I am not a doctor, Maam. I do not diagnose." I know that these are the words that I led with because I am like Chuckie, the wind-up monkey. I always say those sentences.

I absolutely make that disclaimer. No one is ever going to come back with: *The pharmacist said, your honor* because this pharmacist did not say and he never will say.

If they want to tell me their symptoms, I'll do my best to direct them to something that I think will probably help. Palliative care, mostly. Don't even hint at diagnosing. It's not your job and there is too much to lose.

One exception for me. I will always sift through a child's hair with my fingers looking for the tell tale signs of lice. The average mother is not only worried, she is mortified. Rich or poor! Lice is a great equalizer. She thinks she is a bad mother. *How could my child be so dirty? I bathe her every night. I wash her hair.*

I always, without fail, lice or not, say, "This does not make you a bad mother."

I usually get only a half smile. I'll touch her arm and offer to look in *her* hair. In the end, I'll go back to the pharmacy and scrub my hands, lice or not. I'm always itching all over.

I don't like it when they tell me that I am better than a doctor. I am not a doctor. I am a pharmacist and the difference is huge. The average doctor knows about the 50 or so drugs that they actually prescribe. That's okay because you and I know enough about the rest.

Rule Number Seventeen
Don't Be A Wimp

In 1964, pharmacy was a second tier medical profession. We were lumped with the lab technicians and physical therapists. We never questioned a doctor on anything. To make the story even more dreadful, we never entertained the possibility that the doctor might need help. It wasn't in our paradigm to scrutinize an MD. We were lightweights. We were *dispensers. Period!* We were not allowed to counsel. But some of us did. I counseled when it was in the best interests of the patient. Some doctors had real problems with a pharmacist telling their patient anything. It was a turf issue.
That was then. This is now.

Most of us worked in drug stores in 1964. The chains were not dominant yet and some of us made a nice living selling merchandise. Prescriptions were just another department. Camera made more profit than prescriptions. When you landed a good line like Estee Lauder or Chanel, you could double or triple your money. A bottle of Chanel No. 5 sold for $5.00. An Armour Thyroid 1 grain sold for $1.99 for 100 tablets.

The cosmetic companies paid the cosmetologist commissions in cash. It came in a white envelope every month. The cosmetologist drove a Chrysler Imperial. The 50 year old career staff pharmacist drove a Chevrolet Bel Air. She ate lunch in the hotel restaurant and had sales people waiting to pay. The pharmacist brought ham and cheese sandwiches wrapped in waxed paper and drank lukewarm coffee from a Thermos.

In 1964, druggists were merchants first. We were institutionalized to be dispensers. We were even forbidden to put the name of the drug on the label.

In the twenty-first century, you know exactly what authorizations you have in exercising your knowledge and experience. If anything lags, my observation is that we are not very aggressive about seeing that the patient comes first. Bad handwriting can delay the delivery of a critical Rx for hours or even days.

Recently, a patient asked me to call the doctor and ask for a change. She wanted APAP/Codeine 30 mg instead of hydrocodone/APAP 10/325. I complied. The patient was in a wheelchair and had to wait so long she fell asleep.

The doctor was clearly exasperated. She was at a party. "I want you to fill the Norco. That's why I wrote it."

"The patient has had bad luck with Norco."

"It's the same thing."

I paused a moment and then said, "You know better than that, doctor. It is NOT the same thing."

Her turn to pause. "Okay. Okay. Change it then."

There are still older doctors out there who live their lives as if they are infallible and never make mistakes. What a burden when someone dies. It is our job to tell them when they make a mistake about drugs.

The young doctors are easy, but the old ones can be difficult. You better come loaded for bear. He is just the school yard bully, mostly bluff and bravado. Just calmly present the facts and he'll get in line. He will eventually understand that you are saving his ass.

For some older doctors, it is just a knee jerk reaction when a lowly pharmacist calls. He has no idea of what you know these days. "I have a reason for writing both of those drugs. Just fill them." He remembers the days when pharmacists could be backed off by bluster.

I've been a pharmacist for 43 years and my knowledge is limited compared to the average PharmD.
But I do know what I know and I put it to use.

Some of us are still much too timid. Some of us are still dispensers and actually resent having to counsel. That kind of thinking will screw the goose.

You can't hide, man/woman. You have a horse in this race whether you like it or not.

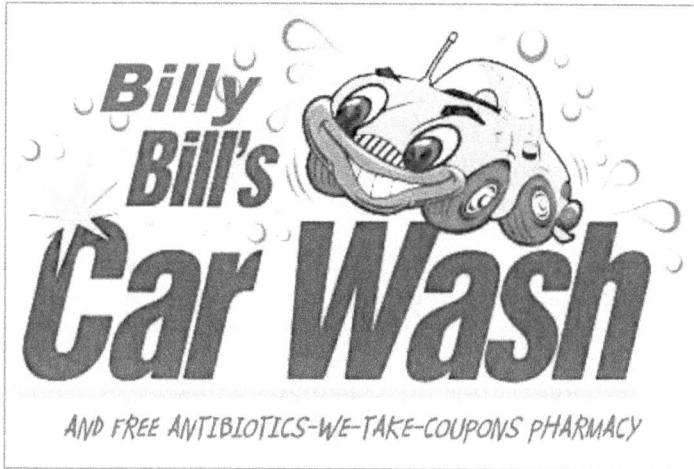

AND FREE ANTIBIOTICS-WE-TAKE-COUPONS PHARMACY

Rule Number Eighteen
The Customer is Always Wrong

Retail Irreverence! *Your store manager still believes that the customer is always right.* Just dare skip the hot wax because you are too busy checking prescriptions for accuracy and your ass will be grass. The litany will be repeated in the *talk* that comes before the dreaded *write-up*.

The 21st Century rule is: *The customer is always wrong in a dispute with the pharmacist.*

You know what you are doing. If anyone is going to correct you, it better be another pharmacist. You are an idiot if you let a non-pharmacist with an Associate's degree from East Insignificant Junior College interfere in any pharmacy business. The good ones will trust that you know what you are doing. You have to set limits with the bad ones.

You are the pharmacist and, according to the old rule, you are supposed to let the customer tell you what is up. I don't think so.

Every single day, a customer will be wrong. "This is not my medicine. My medicine is round and blue. This medicine is

yellow and football shaped."

An accurate observation. This is a young doctor. He wrote *new strength* on the prescription. You explain that the doctor lowered the strength and the patient has a fit.

"I don't want the lower strength, Missy. I want the strong ones. I need the strong ones. My pain is strong. You made a mistake, Missy and I'm gonna report you."

He is right in your face and you back up quickly. You are five foot and 100 pounds. He makes two of you. You lock the Dutch door into the pharmacy and he starts rattling it.

He clearly wants a fight, so you pick up his prescription and walk away quickly.

"Gimme those pills."

"You haven't paid for them yet and these are exactly what the doctor ordered." By now the assistant manager of the store has shown up. He gives you a black stare and starts to kiss some customer ass.

First, you do not need to put up with this. Second, you are very busy. Third, you are putting this Rx in the locked narcotic cabinet until you can call the doctor when the second pharmacist comes in. You have no time this morning and no time for this idiot assistant manager.

"Leave the pharmacy," you tell him. "We are busy and we don't need you sucking up to this guy who is trying to work me."

That afternoon, the non-pharmacist store manager calls you to the office for *the talk*. He has papers on his desk. He is writing notes.

Why are you even expected to explain yourself to this Bozo. You are the pharmacist and you are competent. Why should you have to put up with being questioned by any non-pharmacist who thinks your job is pushing a commodity?

"I want you to write a letter of apology to Mister Kuchinskhy. I want to read it before you mail it. A copy will go in your file." He leans back in his chair and folds his hands across his ample stomach. The sunuvabitch gives you a *gotcha* grin.

You smile back and he clears his throat.

"I'm not writing anything except a serious letter of complaint to the Director of Pharmacy with a copy to the State Board."

"You *have* to write the letter."

"I will."

"Which one?"

"The letter of complaint."

"Complain about what?"

"You."

"But.. Ah...You have to.."

"I'm going to tell him that I can go to work for CostLess Drug next week. Any more crap from you and I'm outta here. Leave me alone."

You are always 5 foot tall right!

Rule Number Nineteen
You Are Not That Important

Did you read about the new doctors who let payment of their student loans lapse? They felt that the country owed them something. They were doctors. Too important to have to pay. They were right. They won't have to pay as long as they stay in West Cornflower, Wyoming, population 505, for the next 5 years at a salary of a teacher. The new teacher is making payments, but she can live anywhere she wants.

Pharmacists may have some *doctor syndrome* issues early in a career, but they usually get over it. An unimpressed, unfriendly and detached demeanor is replaced by a true pharmacist. Relaxed, welcoming, gracious and responsive. It just takes some time. Pharmacists soon give up doing what doctors do, which is keep *ordinary* people at a distance. Frankly, doctors can often look at them as objects. Simple cash cows.

Grown up pharmacists are not like that. We have a *cachet* all of our own. Our prestige comes from being nice people. We are approachable and have very little snob appeal. We are not like doctors or lawyers or even some dentists. People like us and that is better than being intimidating.

In 1968, I had doctor syndrome so bad that my butt was tight. I hung around with dentists and doctors and lawyers. I learned my lessons at a New Years Eve party that year.

This was the manna of life for my first wife, Donna. I think that she actually did sleep with the first dentist to go after her. I was pretty naïve. I couldn't help myself because I had all of the worst symptoms of *doctor syndrome*.

Here I was, all decked out in my grey mohair slacks, white turtleneck and a double-breasted blue blazer. I had just worn the same outfit to a concert at the Carousel Ballroom on Market Street a few weeks earlier. My wife disappeared for three hours. She liked Hippies with dope as much as she liked doctors with martinis.

I was satisfied with the entertainment. The Holding Company with Janis Joplin before she was *Janis*. There were a handful of people dressed like Donna and I. The rest were hippies. Everyone was happy. The marijuana smoke was so thick that no one cared when the bar closed. I vowed to wear jeans the next time.

New Years Eve. I was standing in a circle of professionals. They were boasting about their investments. Condominiums were a favorite. They had units at places like Zephyr Cove at Lake Tahoe and Sea Ranch on the coast. I nodded like I was one of them. I had my eye on Donna. She had been slow dancing with one particular dentist for a long time.

Then this. "What are you doing with your money, Plagakis?"

"Ahm humfm Ah…."

That was the end of protracted *doctor syndrome*. I let my hair grow over my ears. I read Hesse's Siddhartha. I quoted Kahlil Gibran. I was free. I was a pharmacist. Donna did not like that as much. Oh well.

I love pharmacists. I have been hanging out with them for most of my life. They are so out-of-the-ordinary when you look at all of the medical professionals.

Notice how pharmacists dress. Rarely do you see them out with shirts and ties because most of them are women. Some of them wear a blouse and skorts to work. Almost all of them wear sneakers or coach's shoes. No *doctor syndrome* here.

Pharmacists relate to each other like real people in real life. Often, they are jeans-wearing, Dr. Pepper-drinking, Subway sandwich-eating while they are working. They tell stories. They laugh.

Everybody gets a shot at *DS* and everybody needs to get over it.

Rule Number Twenty
Live With Intention, Walk to the Edge
Practice Wellness
Continue to Learn and
Play With Abandon

Do what you love! I know your dreams and how fragile they are. I still have mine. Some have filtered away in the stream of life while others came true unexpectedly. I have no regrets, nor should you. In 1999, Victoria came true when I had lost all hope.

Some of you have given up and are settling. You have a good life, a fine life, but there is a steady gnawing discomfort about what could have been. Big paychecks can't help. You can't buy what you want so dearly   The vision that lit you up is not for sale. You set it aside for later and then *later happened without it.*

How did it get away? You made a world-class mistake by not sharing your dreams. Sharing makes them real in the universe. It sets powerful influences to work in your favor. While you sleep, celestial alliances come to your aid.   Too hippie-dippy for you? Well, I *am* hippie-dippy. I've been that way for decades and I'm not going to be any other way just because people are watching me. I have earned my life.

Dreams are magic and you are the magician. You pull them out of the hat, let everyone see them and then just let them be. Let them mature and blossom out in the ether dimension, just a little bit out of the way from your earthly senses.

For me, a conjurer who learned by taking baby steps, my castles in the sky did not come true for decades after I had first, timidly let them come out and see the light of the sun. I could no longer hide them. Courage came from deep inside (perhaps outside) and my most hidden secret longings filtered out where all in the heavens and on Earth could see them.

This wasn't easy because we are discouraged to believe in the unexplainable. We are told that we get no reward from anything other than hard work. The puritanical bugaboo that all good things take *time* and *effort* will stop a dream from blossoming every time.

My dreams grew all by themselves without any work by me. One day, I realized that one of them had come true, then another. It is always a surprise. What a thrill.

I believe in dragons and angels. I live my life like there are stubborn heavenly forces working for me. I live like a 1960s flower child because I am one. I *do* believe in miracles. They have come to be and all I had to do was dream them.

Dreamless people laughed at me. That is a real danger. You just have to not care. They don't have dreams, so how dare you?

Don't despair. Life without dreams is icy. Some of you will hold on no matter what. Your force of will is simply a flimsy shroud when you compare it with the magic in your favor.

Who are these seekers who are living their lives as I have lived mine? Who are these professionals who have no doubt that pharmacy is *what they do* and not *who they are?*

Who are these sisters and brothers who have learned to laugh right along with me? Who are my partners in the cosmic dance? Who hears the music that I hear?

They are the ones who learned to navigate lightly in a professional world that can crush you and rob you of your humanity and your ability to do magic.

Pharmacy is no different than any other scholarly discipline. The nature of any profession is to homogenize and consume every practitioner and to rip the spirituality right out of her.

That is what happens when you define yourself as what you do. Be careful out there!

The End

Jim Plagakis, RPh
3100 75th Street #14
Galveston, TX 77551
jpgakis@hotmail.com
Cell: 409-739-7403

Resources You Might Find Interesting

http://www.jimplagakis.com

You can find a few years worth of Jim Plagakis' column JP at Large at the Drug Topics Website

http://www.drugtopics.com

You can also subscribe to have the digital edition of Drug Topics delivered directly to your in-box every month.

http://advanstar.replycentral.com/?PID=301&V=DIGI

JP at Large Up to Date is a compilation of the first 169 of Jim Plagakis' columns that first appeared in Drug Topics Magazine in January, 1989. It is published by Advanstar and is available here:

http://www.industrymatter.com/jpatlarge.aspx